D0331022

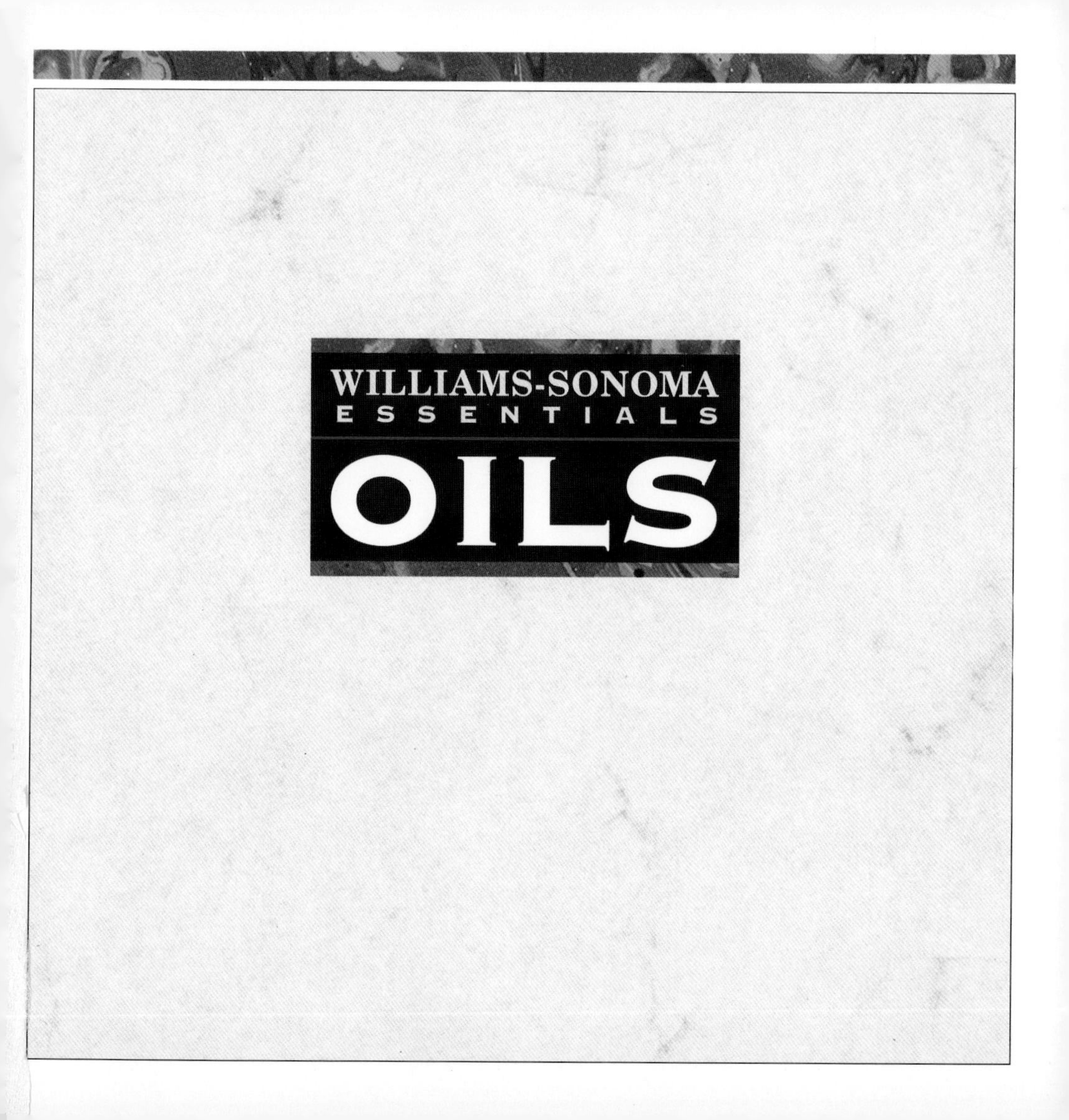
WILLIAMS-SONOMA
ESSENTIALS
OILS

WILLIAMS-SONOMA
ESSENTIALS
OILS
GENERAL EDITOR
CHUCK WILLIAMS
RECIPES
ANN CREBER
PHOTOGRAPHY
PHILIP SALAVERRY

WILLIAMS-SONOMA
Founder: Chuck Williams

WELDON OWEN INC.
President: John Owen
Publisher: Wendely Harvey
Managing Editor: Jill Fox
Consulting Editor: Norman Kolpas
Design & Illustration: Brenda Duke
Editorial & Design Assistant:
Marguerite Ozburn
Recipe Assistants: Cathie Graham,
Janet Lodge
Proofreader: Meredith Phillips
Indexer: ALTA Indexing Service
Production: Stephanie Sherman,
James Obata
Photography Assistant:
David Williams
Food Stylist: Bruce Yim
Food Stylist Assistants:
Barbara Bragle, Tim Scott
Prop Stylist: Amy Glenn

Production by:
Mandarin Offset, Hong Kong
Printed in China

A Weldon Owen Production

WILLIAMS-SONOMA ESSENTIALS
Conceived and produced by:
Weldon Owen Inc.
814 Montgomery Street
San Francisco, CA 94133
Phone number: (415) 291-0100
Fax number: (415) 291-8841

In collaboration with:
Williams-Sonoma
100 North Point Street
San Francisco, CA 94133

Library of Congress
Cataloging-in-Publication Data:
Creber, Ann.
Oils / recipes. Ann Creber ;
photography, Philip Salaverry.
p. cm. -- (Williams-Sonoma
essentials)
ISBN: 1-875137-21-1
1. Oils and fats, Edible. 2. Cookery.
I. Title. II. Series.
TX407.034C74 1994
641.6--dc20 94-4155
CIP

ACKNOWLEDGMENTS
The publishers would like to thank the following people and organizations for their assistance in lending props for photography: Paul Bauer, Inc.; Gail Cohen; B.R. Cohn Vineyard; Cyclamen Studios, Julie Sanders, Designer; Luna Garcia; Judy Goldsmith and Bernie Carrasco, J. Goldsmith Antiques; Stephanie Greenleigh; Sandra Griswold; Missy Hamilton-Backgrounds; Glenda Jordan; Mimi Koch-Backgrounds; Merna Oeberst; Bob Pool; and Zia Houseworks.

WEIGHTS AND MEASURES

All recipes include customary U.S. and metric measurements. The metric conversions are based on a standard developed for these books and have been rounded off. The actual weights may vary. Unless otherwise stated, the recipes were designed for medium-sized fruits and vegetables.

CONTENT

OIL ESSENTIALS

The oils celebrated in this book are prized in culinary circles for the distinctive character they add to recipes. Edible oils are pressed from a range of sources: fruit, especially olive; nuts, including walnut, almond and hazelnut; and vegetables, such as safflower, grapeseed, sunflower and sesame. Oils can be infused with herbs, fruit and spices as well for an even broader range of flavors.

The recipes in this book are organized by the type of oil they feature. Keep a well-stocked pantry of oils for cooking this exciting selection of dishes and creating your own. All oils will keep 6–12 months when stored in airtight containers in a cool, dark place.

OLIVE OIL

With its rich flavor and range of culinary uses, olive oil deserves its reputation as queen of edible oils. Also, medical studies show that monounsaturated fat, which is found in olive oil, may reduce the risk of heart disease, cancer and diabetes.

Olive oil is produced throughout the Mediterranean and in California. Olive oils vary in character and color, but most are made the same way. Olives are crushed in a press and the juice is pumped into tanks where the oil separates and is drawn off. The best oil is from the first pressing, described on some labels as "cold-pressed."

Olive oils are also identified by their acidity. The appropriate choice depends on the recipe. For additional flavor, olive oil can be infused with herbs, fruits and spices.

Extra-Virgin Olive Oil

The finest olive oil, extra-virgin, extracted in a single cold pressing, is the least acidic. Green to gold in color, it has a full-bodied flavor and aroma. Its low acidity—1% or less—makes it smooth on the palate when used in uncooked dishes or added to hot dishes at the end of cooking.

Virgin Olive Oil

Extracted in a single cold pressing, this oil has similar color and aroma to extra-virgin but its higher acidity —1 to 3%—yields a sharper taste best suited to cooked dishes.

Pure Olive Oil

Blending cold-pressed and refined olive oils forms a product more acidic than virgin olive oil and with a taste and color that is not quite as full-bodied as extra-virgin. It may be labeled simply olive oil. Pure olive oil works well in cooked dishes.

Light Olive Oil

Olive oil designated as light is pure olive oil blended to be extra mild. With none of the olive's aroma or flavor, it is ideal for high-temperature cooking or in place of vegetable oils for baking.

Potato and Parsley Frittata

Extra-virgin olive oil adds richness to this flavorful Italian version of the hearty omelet.

- 1 *small leek*
- 2 *tablespoons extra-virgin olive oil*
- ½ *cup (¾ oz / 20 g) finely chopped parsley*
- ¼ *cup (1 oz / 30 g) finely grated Parmesan cheese*
- ⅓ *teaspoon freshly ground pepper*
- 8 *eggs, lightly beaten*
  Salt
- ¾ *lb (375 g) potatoes, peeled, boiled and diced*
- 2 *tablespoons unsalted butter*

Wash the leek very well, discard the coarse green leaves and thinly slice the white part and a small amount of the lower green leaves. In a saucepan heat the olive oil, add the leek and cook over low heat, 3–4 minutes. Remove from heat and set aside to cool. Stir in the parsley, cheese, pepper and eggs. Salt to taste. Gently stir in the potatoes. Preheat a broiler (griller). In an 8-inch (20-cm) frying pan heat the butter. When the butter begins to foam, pour in the leek mixture. Cook over low heat until almost set, 10–15 minutes. Place under the broiler and cook until set, about 1 minute. To serve, cut into wedges for a main course or small squares for appetizers.

Serves 8

Peppered Potato Wedges

These savory morsels are a staple of *tapas,* a Spanish meal made from many small dishes washed down with plenty of sangria.

- 4 *tablespoons virgin olive oil*
- 2 *large potatoes, peeled and cut into thick wedges*
- ¾ *teaspoon coarse salt*
- 1 *teaspoon sweet paprika*
- ⅓ *teaspoon cayenne pepper*

Preheat an oven to 450°F (230°C). Pour 2 tablespoons of the olive oil into a small ovenproof dish. Put in the oven and heat until a light haze forms on the oil. Arrange the potatoes in a single layer in the hot oil. Trickle on the remaining oil, sprinkle with the salt and return to the oven. Bake until the potatoes are crisp outside and the flesh is tender, 30–35 minutes, turning over after 15 minutes. Transfer to paper towels to drain briefly. To serve, place onto a serving platter. Sprinkle with the paprika and cayenne.

Serves 2–12

Salmon Carpaccio

If necessary, place the fresh raw salmon slices between sheets of plastic wrap and pound very gently to make wafer-thin slices. Serve with buttered brown bread.

5 *large ripe tomatoes, peeled, seeded and finely chopped*
1 *small red onion, very finely chopped*
3 *teaspoons finely chopped chives*
Salt
2 *cups (2 oz / 60 g) mesclun or other salad greens*
6 *very thin slices raw salmon*
6 *very thin slices smoked salmon*
18 *very small Ligurian or Niçoise olives*
4 *tablespoons extra-virgin olive oil*

In a medium bowl mix together the tomatoes, onion and chives. Salt to taste. To serve, arrange the mesclun on individual plates. Top with slices of each type of salmon. Spoon portions of the tomato mixture onto the fish. Place 3 olives onto each plate. Drizzle with the olive oil.

Serves 6

Mussels Plaki

Serve these Greek-style shellfish treasures cold as an hors d'oeuvre or as a light main course.

48 *large mussels*
2 *cups (16 fl oz/500 ml) water*
Black peppercorns
½ *cup (4 fl oz/125 ml) pure olive oil*
2 *onions, chopped*
1 *carrot, peeled and diced*
1 *celery stalk, finely diced*
1 *potato, diced*
2 *garlic cloves, finely chopped*
2 *teaspoons superfine (castor) sugar*
3 *tomatoes, seeded and chopped*
3 *tablespoons finely chopped fresh parsley*
2 *tablespoons tomato purée*
Salt and freshly ground pepper

Prepare the mussels by scrubbing shells well and removing beards. Discard any mussels that are open and do not close to the touch. In a large pot combine the mussels, water and a few peppercorns. Cover, bring to a boil and cook 5–7 minutes. Discard any mussels that do not open after cooking. Strain the liquid through cheesecloth and reserve. When the mussels are cool enough to handle, remove them from the shells. In a large frying pan heat the olive oil and sauté the onions until golden. Add the carrot and cook 2 minutes. Add the celery, potato, garlic, sugar, tomatoes, parsley, tomato purée and reserved liquid. Salt and pepper to taste. Cook uncovered over very low heat, stirring occasionally, until the vegetables are tender and the sauce is rich and reduced, about 15 minutes. Add a few drops of water to prevent burning, if necessary. Add the mussels and simmer 5 minutes. Cool to room temperature and then chill. To serve, return the mussels to their half-shells and arrange on individual plates.

Serves 4–6

Chicken Roulades

This colorful dish is delicious served with roasted potatoes.

4 *chicken breast halves, about 7 oz (200 g) each, boned and skinned*
½ *cup (4 fl oz / 125 ml) Pesto Basil Oil (recipe on page 23) or light olive oil*

Preheat an oven to 350°F (180°C). Lightly oil a baking sheet. Place each chicken breast between two sheets of plastic wrap and, using a rolling pin, gently flatten to about ½-inch (12-mm) thick. Spread an equal portion of the Pesto onto each fillet. Roll each fillet to enclose the Pesto and fasten with a toothpick. Lightly brush the roulades all over with the Basil Oil or light olive oil and put onto the baking sheet. Bake until the breasts are opaque throughout when pierced with a knife, about 25 minutes. To serve, slice and arrange on individual plates.

Pesto

3 *cups (3 oz / 90 g) fresh basil leaves, packed*
1 *garlic clove, chopped*
1 *tablespoon pine nuts*
½ *cup (2 oz / 60 g) freshly grated Parmesan cheese*
5 *tablespoons (3 fl oz / 80 ml) virgin olive oil*

In the work bowl of a food processor or blender combine the basil, garlic, pine nuts and cheese. Add a dash of the olive oil and pulse two or three times. With the motor running, slowly pour in the remaining oil. Process to a thick, creamy consistency.

Serves 4

PROVENÇAL POT ROAST

Herbes de Provence—a blend of dried thyme, lavender, summer savory, basil and rosemary—ensures that the authentic flavors are retained in this dish from the south of France. Serve over boiled noodles, in the traditional way.

- 2 *lb (1 kg) rump roast*
- 1 *cup (8 fl oz/250 ml) red wine*
- ¼ *cup (2 fl oz/60 ml) extra-virgin olive oil*
- 2 *garlic cloves, peeled and chopped*
- 1 *teaspoon Herbes de Provence*
- ½ *teaspoon peppercorns*
- ¼ *cup (2 fl oz/60 ml) vegetable oil*
- 1 *onion, chopped*
- 1 *cup (6 oz/185 g) ripe tomatoes, peeled and chopped*
- 2 *teaspoons orange zest*
- 2 *anchovy fillets, chopped*
- 2 *tablespoons black olives, sliced*
- *Salt*

Trim all the fat from the meat. In a large bowl combine the wine, olive oil, garlic, Herbes de Provence and peppercorns to form a marinade. Add the meat and coat evenly. Cover lightly and refrigerate 1–2 days, turning occasionally. Preheat an oven to 325°F (165°C). In a heavy cast-iron casserole heat the vegetable oil. Add the beef and sear all over. Pour in the marinade, onion, tomatoes, orange zest and anchovies and bring to a boil. Cover and cook until the meat is tender, 2½–3 hours. Add the olives during the last 10 minutes of cooking. Salt to taste. To serve, carve the beef into thick slices. Top with the juices from the casserole.

Serves 4–6

INFUSED OIL

Combining olive oil with distinct flavoring ingredients results in special, aromatic oils for seasoning food.

Use virgin or light olive oils as a base for infused oils. Extra-virgin olive oil is too strong to be infused and will overwhelm the flavors added to it.

Infuse oil with well-washed herbs, citrus peels and spices. Heat—whether from the stovetop or sun—induces the flavors to marry. In most cases you'll need to strain the infused elements out of the oil prior to sealing in bottles. Adding fresh elements after straining makes for a stunning presentation but does little to the flavor. Remove these elements after a few weeks as they can get moldy, ruining the wonderful oil.

Rosemary Oil

2 cups (16 fl oz/500 ml) virgin olive oil
½ cup (1 oz/20 g) fresh rosemary leaves or
1 tablespoon dried rosemary, crushed
Fresh rosemary sprigs

In a saucepan over low heat, warm the oil 5 minutes. Put the fresh or dried rosemary leaves into a glass jar. Pour in the warmed oil and seal. Leave in a warm spot until flavors meld, about 1 week. Strain the oil into sterilized presentation bottles, add the rosemary sprigs, if desired, and seal. Makes 2 cups (16 fl oz/500 ml).

Peppercorn and Chili Oil

6 fresh or dried red chilies
⅓ cup (2 oz / 60 g) mixed black, white and pink peppercorns
2 cups (16 fl oz / 500 ml) virgin olive oil

In a glass jar combine the chilies, peppercorns and oil and seal. Leave in a warm spot until flavors meld, about 1 week. Strain the oil into sterilized presentation bottles and seal. Makes 2 cups (16 fl oz / 500 ml).

Chili Oil

2 cups (16 fl oz / 500 ml) virgin olive oil
12 fresh or dried red chilies

In a saucepan over low heat, warm the oil 5 minutes. Put 8 of the chilies into a glass jar, pour in the warmed oil and seal. Leave in a warm spot until flavors meld, about 1 week. Strain the oil into sterilized presentation bottles, add the remaining chilies, if desired, and seal. Makes 2 cups (16 fl oz / 500 ml). Note: Always wash hands after handling any chilies.

SUN-DRIED TOMATO OIL

- 2 *cups (16 fl oz/500 ml) virgin olive oil*
- ½ *cup (4 oz/125 g) sun-dried tomatoes, chopped*
- ½ *teaspoon black peppercorns*
- 1 *fresh thyme sprig*

In a saucepan over low heat, warm the oil 5 minutes. Add the tomatoes and peppercorns and cook 5 minutes. Remove from heat and cool until cold. Pour into sterilized presentation bottles, add the thyme and seal. Leave in a warm spot until flavors meld, about 2 weeks. This oil does not need to be strained prior to use. Makes 2 cups (16 fl oz/500 ml).

LEMON THYME OIL

- 2 *cups (16 fl oz/500 ml) virgin olive oil*
- ¾ *cup (1 oz/30 g) fresh lemon thyme leaves*
- ½ *teaspoon black peppercorns*
- 2 *twists of lemon peel*

In a saucepan over low heat, warm the oil 5 minutes. Put the lemon thyme into a glass jar. Pour the warmed oil over the herbs, add the peppercorns and seal. Leave in a warm spot until flavors meld, about 2 weeks. Strain the oil into sterilized presentation bottles, add the lemon peel twists, if desired, and seal. Makes 2 cups (16 fl oz/500 ml).

CITRUS OIL

2 *cups (16 fl oz/500 ml) light olive oil*
2 *twists of lime peel*
2 *twists of lemon peel*
2 *twists of orange peel*
½ *teaspoon peppercorns*

In a saucepan over low heat, warm the oil 5 minutes. Put the citrus peels into sterilized presentation bottles. Pour the warmed oil over peels, add the peppercorns and seal. Leave in a warm spot until flavors meld, about 2 weeks. This oil does not need to be strained prior to use. Makes 2 cups (16 fl oz/500 ml).

TARRAGON OIL

2 *cups (16 fl oz/500 ml) virgin olive oil*
⅓ *cup (½ oz/13 g) crushed fresh tarragon leaves or 4 teaspoons dried tarragon*
1 *fresh bay leaf*

In a saucepan over low heat, warm the olive oil 5 minutes. Put the tarragon and bay leaf into a glass jar. Pour the warmed oil over the herbs and seal. Leave in a warm spot until flavors meld, about 2 weeks. Strain the oil into sterilized presentation bottles and seal. Makes 2 cups (16 fl oz/500 ml).

BASIL OIL

¾ *cup (1 oz/30 g) crushed fresh basil leaves or 1 tablespoon dried basil leaves*
2 *cups (16 fl oz/500 ml) light olive oil*

In a glass jar combine the basil and olive oil and seal. Leave in a warm spot until flavors meld, about 2 weeks. Strain the oil into sterilized presentation bottles and seal. Makes 2 cups (16 fl oz/500 ml).

Chilled Lettuce and Onion Soup

An excellent way to use the coarse outer lettuce leaves or to solve the problem of excess lettuce when it all matures at once in the garden.

- 2 *tablespoons Tarragon Oil (recipe on page 23)*
- 1 *large iceberg lettuce, or about ½ lb (250 g) lettuce leaves, shredded*
- 1 *small white onion, chopped*
- 1 *celery stalk, chopped*
- 6 *green (spring) onions, chopped*
- 3 *tablespoons coarsely chopped fresh parsley*
- 7 *cups (56 fl oz/1.75 l) chicken stock*
- 1 *teaspoon salt*
- ⅓ *teaspoon freshly ground pepper*
- ¾ *cup (6 fl oz/180 ml) sour cream*
- 1 *tablespoon chopped fresh tarragon*

In a large pot, heat the Tarragon Oil. Reserve one fourth of the lettuce and sauté the remainder for 3 minutes. Add the white onion, celery, two thirds of the green onions and parsley and cook another 3 minutes. Add the stock, salt and pepper and simmer 10 minutes. Transfer to the work bowl of a food processor or blender and process until smooth. Stir in the remaining lettuce and green onion and chill. To serve, ladle into individual bowls. Swirl portions of the sour cream into each bowl. Garnish with the tarragon.

Serves 6–8

TABBOULEH

Serve this nutritious Middle Eastern salad in cabbage leaf cups or with toasted pita bread.

- 2 *cups (12 oz / 360 g) bulgur (cracked wheat)*
- 1 *cup (1½ oz / 40 g)* each *finely chopped fresh parsley and basil*
- 3 *large tomatoes, finely chopped*
- 1 *small onion, finely chopped*
- 3 *tablespoons fresh lemon juice*
- 5 *tablespoons (3 fl oz / 80 ml) Basil Oil (recipe on page 23)*
- *Salt and freshly ground pepper*

Put the bulgur into a bowl and pour on enough boiling water to cover the grain. Soak 1–2 hours. Drain the bulgur into a towel, squeezing out as much liquid as possible. Put the bulgur into a bowl and fluff with fork or fingers. Add the parsley, basil, tomatoes, onion, lemon juice and Basil Oil. Salt and pepper to taste and mix well. Cover and chill until serving time.

Serves 6

Avocado Salsa Salad

Serve this beautiful, fresh salad for a summer luncheon party.

- 2 *ripe avocados, pitted, peeled and thinly sliced*
- 1 *lime*
- 4 *ripe tomatoes, seeded and finely chopped*
- 2 *small red onions, thinly sliced stem to root*
- 1½ *tablespoons Citrus Oil (recipe on page 23)*
- 1 *tablespoon finely chopped fresh cilantro (coriander) leaves*
- *Salt and freshly ground pepper*

Arrange the avocado slices on individual serving plates. Juice one half of the lime. Slice the remaining half lime into 4 pieces. Brush the avocados with the lime juice. In a bowl combine the tomatoes, onions, Citrus Oil and cilantro. Salt and pepper to taste. To serve, spoon the mixture over the avocado slices. Garnish with the lime slices. Serve immediately.

Serves 4

Basque Baguette

This vibrantly flavored hors d'oeuvre, from a tiny restaurant in the Pyrenees, is a wonderful choice to serve at a cocktail party.

- 3 *large red bell peppers (capsicums)*
- ½ *lb (8 oz / 250 g) ripe tomatoes, sliced*
- 1 *teaspoon superfine (castor) sugar*
- 1 *baguette cut into ½-inch (1.25-cm) thick slices*
- 2 *teaspoons tomato purée*
- 1 *teaspoon sweet paprika*
- 1 *garlic clove, crushed*
- 2 *tablespoons red wine vinegar*
- *Salt and freshly ground pepper*
- 6 *tablespoons (3 fl oz / 90 ml) Sun-Dried Tomato Oil (recipe on page 22)*

To roast the peppers, preheat an oven to 400°F (200°C). Cut peppers in half lengthwise. Remove ribs and seeds. Place cut-side down onto an ungreased baking sheet, flattening with hand. Roast until blistered and blackened, about 40 minutes, turning peppers several times during cooking. Remove the peppers from the oven, place into a paper bag, seal and let stand 20 minutes. Peel and cut peppers into strips. Dust the tomato slices lightly with the sugar. Toast the bread slices. In a bowl mix together the tomato purée, paprika, garlic and vinegar. Salt and pepper to taste. Whisk in the Sun-Dried Tomato Oil, pouring in a thin drizzle. To serve, place the pepper strips and a tomato slice on each bread slice. Top each with a tablespoon of the purée mixture.

Serves 4–6

Gratin of Potatoes with Anchovies

Rosemary Oil and fresh basil gently scent this potato casserole. The anchovy can be left out, if desired. Choose a baking dish that can go from the oven to the table.

- 8 *anchovy fillets*
- ¼ *cup (2 fl oz / 60 ml) milk*
- 1 *garlic clove, halved*
- ⅓ *cup (3 fl oz / 80 ml) Rosemary Oil (recipe on page 20)*
- 4 *lb (2 kg) russet potatoes, peeled, thinly sliced and soaked in cold water*
- 1 *cup (1½ oz / 40 g) chopped fresh basil leaves*
- *Salt and freshly ground pepper*
- 2 *cups (16 fl oz / 500 ml) rich chicken stock*
- ½ *cup (2 oz / 62 g) grated Swiss cheese*

To reduce salt, soak the anchovy in the milk for 20 minutes, drain and discard milk. Preheat an oven to 400°F (200°C). Rub base and sides of a 12-inch (30-cm) baking dish with the garlic and brush all over with the Rosemary Oil. Drain the potatoes and pat dry. In the baking dish arrange layers of the potatoes, anchovies and basil until all are used. Salt and pepper to taste. Pour in the stock. Bake until potatoes are tender and liquid is absorbed, 1–1¼ hours. Top with the cheese and bake an additional 10 minutes. To serve, cool several minutes before spooning onto individual plates.

Serves 6–8

GREEN FETTUCINE WITH COLD TOMATO SAUCE

Basil and tomatoes grow well together in the garden. They taste great together in this cold sauce as well.

- 4 *large ripe tomatoes, peeled and chopped*
- 1 *small white onion, finely chopped*
- 1 *garlic clove, finely chopped*
- 2 *tablespoons finely chopped fresh basil*
- *Salt and freshly ground pepper*
- 2 *tablespoons Basil Oil (recipe on page 23)*
- ¾ *lb (12 oz/375 g) spinach fettucine*
- ½ *cup (2 oz/65 g) grated Parmesan cheese*

In a bowl combine the tomatoes, onion, garlic and basil. Salt and pepper to taste. Stir in the Basil Oil. Fill a large pot three-quarters full with water. Bring to a boil, add the fettucine and cook until al dente, about 8 minutes for dried pasta, 2 minutes for fresh pasta. Drain. To serve, spoon the sauce over the hot fettucine and sprinkle with the cheese.

Serves 4

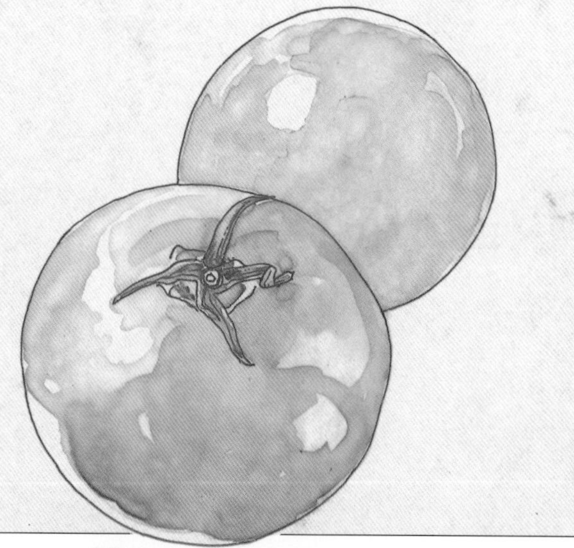

SANTA MARGHERITA CIOPPINO

Not quite a stew, but more than a soup, this fragrant dish is wonderful for entertaining. Serve with crusty baguettes, slathered with butter.

- ¼ *cup (2 fl oz / 60 ml) Basil Oil* *(recipe on page 23)*
- 1 *medium onion, chopped*
- 1 *garlic clove, minced*
- 1 *lb (16 oz / 500 g) ripe tomatoes, peeled and chopped (retain juice)*
- ½ *teaspoon salt*
- ⅓ *teaspoon freshly ground pepper*
- 1 *tablespoon shredded fresh basil leaves*
- 4 *cups (32 fl oz / 1 l) water*
- 1 *cup (8 fl oz / 250 ml) dry white wine*
- 1 *lb (16 oz / 500 g) white fish fillets, cut into chunks*
- ½ *lb (8 oz / 250 g) raw shrimp (prawns), shelled and deveined*
- ¼ *cup (⅜ oz / 10 g) finely chopped fresh parsley*

In a large pot over moderate heat warm the Basil Oil. Add the onion and garlic and cook until the onion is tender, 8–10 minutes. Add the tomatoes and their juice, salt, pepper and basil. Add the water, increase heat and bring to a boil. Reduce heat, cover pot and simmer 30 minutes, stirring often. Add the wine and fish. Return to a boil, reduce heat and simmer, covered, 5 minutes, stirring occasionally. Add the shrimp, cover and cook until the shrimp turn pink, about 5 minutes. To serve, ladle into individual bowls. Garnish with the parsley.

Serves 4–6

Mexican Fish

Rosemary Oil, citrus juice and cilantro add traditional Mexican flavor to any white fish used for this grill recipe.

- 1 *cup (8 fl oz/250 ml) Rosemary Oil (recipe on page 20)*
- 1 *cup (8 fl oz/250 ml) fresh lime or lemon juice*
- 2 *fresh red chilies, stemmed, seeded and sliced*
- 2 *tablespoons chopped fresh cilantro (coriander) leaves*
- *Salt and freshly ground pepper*
- 2 *lb (1 kg) thick white fish fillets*
- *Fresh rosemary sprigs*

In a bowl combine the Rosemary Oil, citrus juice, chilies and cilantro. Salt and pepper to taste. In a ceramic or glass dish arrange the fish and cover with the Rosemary Oil mixture. Cover and chill 15 minutes. Prepare a fire in a barbecue grill. Grill the fish fillets, basting often with the marinade, until the flesh is opaque, about 5 minutes per side. To serve, transfer fish to individual plates. Top with any remaining marinade. Garnish with the rosemary sprigs.

Serves 4

Chili Shrimp

Serve these spicy shellfish with plenty of crusty bread for dipping into the flavorful Chili Oil.

- *1½ lb (750 g) raw shrimp (prawns), peeled and deveined*
- *1 cup (8 fl oz/250 ml) Chili Oil (recipe on page 21)*
- *3 garlic cloves, peeled and sliced*
- *1 fresh or dried red chili, seeded and finely sliced*
- *½ teaspoon salt*

Rinse the shrimp and pat dry. In a deep frying pan heat the Chili Oil. Add the garlic, chili, salt and shrimp. Toss over high heat until the shrimp turn pink, about 5 minutes. (It is essential that the garlic not be allowed to burn as this causes unpleasant bitterness.) To serve, transfer the shrimp to a serving platter. Pour the Chili Oil mixture into a separate bowl to use for dipping.

Serves 4

Grilled Salmon in Lemon Sauce

Serve this flavorful fish with Warm Potato Salad (recipe on page 49).

- 6 *8 oz (250 g), 1-inch (2.5-cm) thick salmon steaks*
- 3 *tablespoons Citrus Oil (recipe on page 23)*
  Salt and freshly ground pepper
- 2 *egg yolks*
- 4 *slices white bread, made into breadcrumbs*
- 2 *tablespoons finely chopped fresh parsley*
- ½ *teaspoon capers, rinsed and drained*
- 1 *garlic clove, peeled and chopped*
- 8 *fresh mint leaves*
- ½ *cup (4 fl oz / 125 ml) Citrus Oil*
  Fresh mint sprigs

Preheat a broiler (griller). Brush the salmon with 3 tablespoons of the Citrus Oil and sprinkle with salt and pepper. In the work bowl of a food processor or blender mix together the egg yolks, breadcrumbs, parsley, capers, garlic and mint. With the motor running, add the Citrus Oil in a very thin stream. Continue to process mixture to a thick consistency and set aside. Broil the salmon until opaque throughout when pierced with a knife, 3 minutes on each side. To serve, place the salmon steaks onto individual plates. Top with a spoonful of the lemon sauce. Garnish with the mint sprigs.

Serves 6

Roast Chicken with Tarragon and Lime

Choose free-range chickens to best enjoy the succulent flavor of Tarragon Oil sharpened with the tang of limes.

2 *3 lb (1.5 kg) roasting chickens*
Salt and freshly ground pepper
6 *garlic cloves, peeled*
4 *limes, quartered*
2 *sprigs fresh tarragon*
2 *tablespoons unsalted butter*
6 *tablespoons (3 fl oz / 90 ml) Tarragon Oil (recipe on page 23)*

Preheat an oven to 425°F (220°C). Season the chicken inside and out with salt and pepper. Inside each chicken place 3 garlic cloves, 2 quarters of lime, a sprig of tarragon and a tablespoon of butter. Rub chickens all over with half of the Tarragon Oil. Put chickens into a baking dish. Roast 15 minutes, pour on remaining oil and baste chicken with pan juices. Roast another 15 minutes, reduce temperature to 360°F (185°C) and roast until golden, another hour. Check chickens every 15–20 minutes and baste with pan juices. To serve, arrange the chickens on a serving platter. Garnish with the remaining lime quarters. Cool chickens 5 minutes before carving.

Serves 6–8

Lamb Medallions

Serve this succulent lamb as a main course with Gratin of Potatoes with Anchovies (recipe on page 30) or on thick slices of French bread as a scrumptious sandwich.

4 6 oz (185 g) lamb tenderloins
1 tablespoon fresh thyme leaves
½ cup (4 fl oz / 125 ml) Lemon Thyme Oil (recipe on page 22)
Salt and freshly ground pepper

Trim the fat from the lamb. Arrange the tenderloins in a shallow baking dish. In a bowl mix together the thyme and Lemon Thyme Oil. Salt and pepper to taste. Stir well and pour over the meat. Cover and refrigerate, turning often, at least 6 hours. Preheat a broiler (griller). Cut the lamb into 1-inch (2.5-cm) thick slices. Broil the meat until well browned, 4–5 minutes per side. To serve, place the lamb onto individual plates.

Serves 4

Garlic and Herb Bread

Serve this flavorful, tasty bread with salads or enjoy it as a snack.

- 1½ *cups (7½ oz / 235 g) all-purpose (plain) flour, sifted*
- ½ *teaspoon salt*
- 2 *teaspoons active dry yeast*
- 4 *tablespoons Basil Oil (recipe on page 23)*
- 2 *garlic cloves, peeled and finely chopped*
- 1 *egg, beaten*
- 4 *oz (125 g) cooked spinach, drained and puréed*
- 1 *tablespoon* each *finely chopped fresh basil and parsley*
- 1 *teaspoon fresh chopped thyme leaves*
- *Freshly ground pepper*
- 1 *tablespoon milk*
- ¼ *cup (30 g) finely grated Parmesan cheese*

Oil an 8-inch (20-cm) tart pan. In a bowl combine the flour, salt, yeast, 1 tablespoon of the Basil Oil and the garlic. In a measuring cup combine half the egg with enough water to equal 5 fl oz (160 ml). Stir the egg mixture into the flour mixture. Mix until a ball forms. Place on a floured board and knead, about 10 minutes. Using a rolling pin, form the dough into a rectangle approximately 9 x 6 inches (23 x 15 cm). Brush the dough with 1 tablespoon of the Basil Oil, leaving a border along one of the long edges. Spread the spinach over the oiled surface, leaving the border clear. Sprinkle on the basil, parsley, thyme and pepper. Lightly moisten the clear edge of dough with a little water, and starting at the opposite long edge, roll up the dough to form a jelly roll. Pinch ends together to seal. Brush all over with the remaining Basil Oil. Cut into 7 equal slices. Arrange slices, cut-side up, in the tart pan. Cover and leave in a warm place until well risen, 1–2 hours. Preheat an oven to 425°F (220°C). In a small bowl combine the remaining egg and milk. Brush onto the dough. Sprinkle on the cheese. Bake until golden, 20–25 minutes. Cool slightly before turning out.

Serves 7

NUT OIL

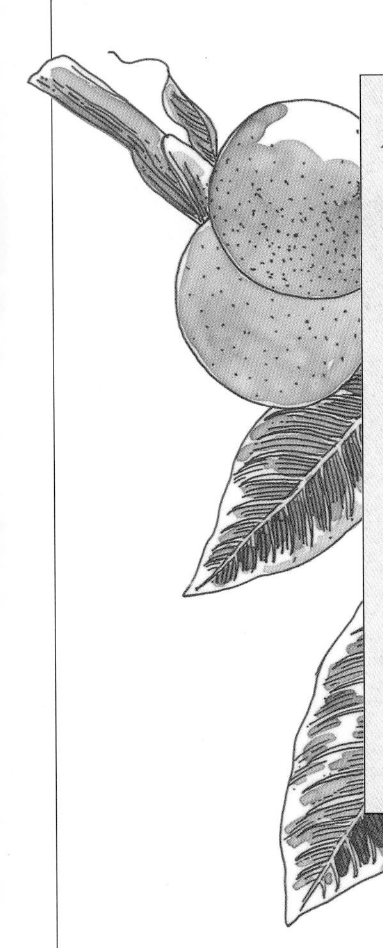

All kinds of nuts yield oils. Two oils stand preeminent, however, when it comes to lending culinary distinction: walnut oil and hazelnut oil.

Cold-pressed from lightly toasted nuts, walnut oil first found its way into European kitchens in the 19th century. Today the United States and France provide most of the world's supply. Along with adding rich, full flavor to salad dressings, use it to flavor grilled or roast meats and baked goods.

More subtle but no less distinctive, hazelnut oil—primarily produced in France—has only been available commercially since 1978. Use it for light sautéing, as a savory salad seasoning or as an ingredient in baked goods.

Nut oils spoil quickly. Purchase small quantities, refrigerate after opening and use within 6 months.

Baked Camembert

Serve small portions of this rich cheese with toast as an appetizer.

- 4 *apples, cored and sliced*
- 2 *pears, cored and sliced*
- *Juice of 2 fresh lemons*
- ¼ *cup (2 fl oz / 60 ml) hazelnut oil*
- ⅓ *cup (¾ oz / 20 g) fresh whole-wheat (wholemeal) breadcrumbs*
- ⅓ *cup (1⅔ oz / 50 g) hazelnuts, toasted, peeled and chopped*
- 1 *lb (500 g) Camembert cheese, cut into 8 wedges*
- 2 *cups (2oz / 60 g) salad greens*

Brush the apples and pears with lemon juice. Pour the hazelnut oil into a dish. In a separate dish, combine the breadcrumbs and nuts. Dip the cheese into the oil, then into the breadcrumb mixture to coat evenly. Arrange the cheese on a baking sheet. Chill for several hours. Preheat an oven to 325°F (165°C). Bake the cheese, uncovered, until the crust is brown and the center begins to run, about 12 minutes. To serve, arrange the greens on a serving platter. Top with the fruit and cheese.

Serves 8

Warm Potato Salad

The combination of walnut oil and fresh walnuts intensifies the rich, earthy flavor of this side dish.

- 1 *lb (16 oz / 500 g) red-skinned potatoes, cut into chunks*
- 3 *green (spring) onions, finely chopped*
- 1 *cup (4 oz / 125 g) shelled walnuts, toasted*
- 1 *tablespoon chopped fresh parsley*
  Salt and freshly ground pepper
- 6 *tablespoons (3 fl oz / 90 ml) walnut oil*
- 2 *tablespoons white wine vinegar*
- 4 *eggs, hard-boiled, peeled and quartered*

Cook the potatoes in boiling water until tender; drain well. In a bowl combine the potatoes and onions and toss gently. Sprinkle in the walnuts and parsley. Salt and pepper to taste. Pour on the walnut oil and vinegar. Add the eggs and toss very gently. Serve warm.

Serves 4

Walnut Beer Bread

Top this beautiful bread with cheese and chutney for a fantastic snack.

- 1½ *cups (12 fl oz / 375 ml) beer, warmed*
- 2 *teaspoons active dry yeast*
- 4½ *cups (22 oz / 700 g) unbleached flour*
- ½ *cup (2 oz / 60 g) rye flour*
- 2 *tablespoons raw sugar*
- 1 *tablespoon salt*
- ½ *cup (4 fl oz / 125 ml) walnut oil, plus oil for brushing*
- 2 *small white onions, finely chopped*
- 1 *cup (4 oz / 125 g) shelled walnuts, coarsely chopped*
- 1 *egg yolk, beaten*
- 2 *teaspoons chopped fresh rosemary*

Oil a baking sheet. Pour one-third of the beer into a small bowl, stir in the yeast and leave for 10 minutes. In a large bowl, mix together the flours, sugar and salt. Make a well in the center and pour in the beer mixture, walnut oil and remaining beer. Mix well. Knead the dough until smooth and elastic, adding more flour, if necessary. Put the dough into a bowl and brush all over with the extra walnut oil. Allow to rise in a warm place until doubled in bulk, up to 1 hour. Punch down the dough; knead in the onions and walnuts. Put the dough onto the baking sheet, form into a round loaf, cover and set aside again to rise. Preheat an oven to 400°F (200°C). When risen, uncover and bake 10 minutes. Reduce heat to 350°F (180°C) and bake 35 minutes. Brush with egg and sprinkle with rosemary. Continue to bake until loaf sounds hollow when rapped with knuckles, 15–20 minutes. Cool on a wire rack before serving.

Makes 1 loaf

VEGETABLE AND SEED OIL

Generally flavorless, oils pressed, refined and sometimes blended from safflower, sunflower and grape seeds are the workhorses of the oil world—used for all-purpose sautéing and frying, in baking and to tone down stronger-flavored oils in dressings and sauces. Grapeseed oil in particular is prized for its high smoking point—446°F (230°C)—that makes it an ideal medium for grilling and deep frying.

Sesame oil bridges two worlds. In its pale refined state, pressed from untoasted seeds, it may be used for general cooking purposes, though it will impart some of the popular seeds' familiar flavor. The darker version, Asian sesame oil, is pressed from toasted seeds and has a distinctive, strong flavor that suffuses any dish to which a few sparing drops are added.

Potatoes Provencale

- 3 *lb (1.5 kg) new potatoes*
- 4 *cups (24 oz / 740 g) cold cooked roast beef, cubed*
- 1 *cup (6 oz / 188 g)* each *cherry tomatoes and chopped green (spring) onions*
- ¼ *cup (⅜ oz / 10 g) chopped fresh parsley*
- 1 *tablespoon chopped capers*
- ¾ *cup (3 oz / 90 g) imported black olives, pitted and halved*
- 1½ *cups (12 fl oz / 375 ml) safflower oil*
- ⅓ *cup (3 fl oz / 80 ml) white wine vinegar*
- 2 *tablespoons chopped fresh basil*
- 1 *teaspoon sugar*
- 2 *garlic cloves, minced*
- 8 *anchovy fillets, chopped*

In a large pot boil the potatoes until tender. Cut into large chunks. In a large bowl combine the potatoes, beef, tomatoes, onions, parsley, capers and olives. In a bowl whisk together the safflower oil, vinegar, basil, sugar, garlic and anchovy. To serve, mix together.

Serves 8

Roast Parsnips

This side dish is a perfect accompaniment to roast beef.

- ¼ *cup (2 fl oz/60 ml) vegetable oil*
- 2 *lb (1 kg) parsnips, peeled and cut lengthwise*
- *Salt and freshly ground pepper*
- 2 *tablespoons sesame seeds, toasted*
- *Fresh parsley, chopped*

Preheat an oven to 450°F (230°C). In a baking dish heat the vegetable oil in the preheating oven 5 minutes. Add the parsnips. Salt and pepper to taste. Roast, turning occasionally, until tender and golden, 35–40 minutes. To serve, sprinkle with the sesame seeds and parsley.

Serves 6–8

Chinese Pearl Pork Balls

Perfect to serve as part of an Asian buffet, the delicate but pervasive flavor of sesame oil contributes an intriguing element to this appetizer.

- 2 *cups (14 oz/440 g) short-grain rice*
- 8 *dried shiitake mushrooms*
- 12 *water chestnuts, finely chopped*
- 1½ *lb (750 g) lean ground pork*
- 2 *large eggs, well beaten*
- 2 *tablespoons light soy sauce*
- 1 *tablespoon chopped or minced fresh ginger*
- 1 *garlic clove, finely chopped*
- ½ *teaspoon salt*

Soak the rice in water to cover 2 hours. Drain, rinse well and drain again. Soak the mushrooms in 1 cup (8 fl oz/250 ml) water until softened, about 30 minutes. Strain, reserve liquid and chop mushrooms finely. In a large bowl combine the mushrooms, water chestnuts, pork, eggs, soy sauce, ginger, garlic and salt, adding enough of the reserved liquid to moisten the mixture. With oiled hands, shape a tablespoonful of the mixture into a ball and roll in the drained rice until well coated. Repeat until all of the mixture is used. Place well apart on an oiled steamer tray and steam until cooked throughout, 1½–2 hours. To serve, transfer to a serving platter. Top with a trickle of the Sesame-Onion Sauce.

Sesame-Onion Sauce

- 3 *tablespoons black rice vinegar or sweet sherry*
- 2 *tablespoons light soy sauce*
- 3 *tablespoons sesame oil*
- 1 *tablespoon green (spring) onion, finely chopped*
- 1 *tablespoon sesame seeds, toasted*

In a small jar combine the vinegar or sherry, soy sauce, sesame oil, onion and sesame seeds. Seal and shake vigorously. Sauce may be made in advance and refrigerated, but serve at room temperature, well shaken.

Makes 50 balls

SKEWERED SCALLOPS

Serve this Asian-inspired dish with steamed rice and bok choy.

- 3 *tablespoons soy sauce*
- 1½ *tablespoons vegetable oil*
- 2 *teaspoons sesame oil*
- 2 *tablespoons mirin or dry sherry*
- 1 *garlic clove, crushed*
- 1 *teaspoon sugar*
- ⅛ *teaspoon wasabi (Japanese horseradish)*
- 1 *lb (500 g) scallops*
- 2 *slices lean, rindless bacon*
- 1 *small red or green bell pepper (capsicum), cut into squares*
- *Juice of 1 fresh lemon*
- 2 *tablespoons finely chopped fresh parsley*

In a bowl whisk together the soy sauce, vegetable oil, sesame oil, mirin or sherry, garlic, sugar and wasabi. Let stand 30 minutes. Trim the scallops and put into a separate bowl. Cover the scallops with the soy sauce mixture and marinate 30 minutes. Preheat a broiler (griller). Thread the scallops, bacon and bell peppers alternately onto 4 soaked bamboo or metal skewers. Broil, brushing occasionally with the marinade, until scallops are cooked through, about 3 minutes per side. To serve, pour on the lemon juice and sprinkle with the parsley.

Serves 4

Sweet Grape Dessert Pizza

Not quite the traditional pizza, but a dessert delight. Choose different jellies for a flavor change.

- 1 *cup (4 oz / 125 g) all-purpose (plain) flour*
- 1 *tablespoon superfine (castor) sugar*
- 2 *teaspoons baking powder*
- 6 *tablespoons (3 fl oz / 90 ml) grapeseed oil, plus oil for brushing*
- ⅓ *cup (3 fl oz / 80 ml) milk*
- ⅓ *cup (3 fl oz / 80 ml) grape jelly*
- 4 *cups seedless red grapes*
- 1 *oz (30 g) confectioners' (icing) sugar*

Preheat an oven to 400°F (200°C). Oil a baking sheet. In a bowl sift together the flour, sugar and baking powder. Make a well in the center and pour in the grapeseed oil and enough milk to form a soft, but not sticky, dough. On a well-floured board, roll the dough into a 9-inch (23-cm) round and put onto the baking sheet. Prick dough all over with a fork, then brush generously all over with the extra grapeseed oil. Bake until lightly browned, 15–20 minutes. Remove from oven and spread about 3 tablespoons of the jelly onto the hot pastry. Distribute the grapes onto the surface and cover with the remaining jelly. Return to oven and bake until the jelly is slightly bubbling and the grapes are hot, 10–15 minutes. To serve, sprinkle with the confectioners' sugar and cut into wedges.

Spiced Carrot and Zucchini Cake

This deliciously moist cake cuts and keeps very well.

- *3½ cups (17 oz / 550 g) all-purpose (plain) flour*
- *1½ teaspoons baking soda (bicarbonate of soda)*
- *1 teaspoon ground cinnamon*
- *1 teaspoon nutmeg*
- *2 cups (1 lb / 500 g) superfine (castor) sugar*
- *4 eggs*
- *1½ cups (12 fl oz / 375 ml) sunflower seed oil*
- *1 cup (6 oz / 185 g) undrained canned crushed pineapple*
- *1 cup (5 oz / 155 g) finely grated carrot*
- *1 cup (5 oz / 155 g) finely grated zucchini (courgette)*
- *1 cup (4 oz / 125 g) shelled pecans, chopped*
- *1 teaspoon vanilla extract*
- *1 tablespoon confectioners' (icing) sugar*

Preheat an oven to 350°F (180°C). Oil a 9-cup (72 fl oz / 2.25 l) bundt pan. In a large bowl sift together the flour, baking soda, cinnamon and nutmeg. Mix in the superfine sugar. Make a well in the center. Add the eggs and sunflower seed oil and beat until well mixed. Add the pineapple, carrot, zucchini, pecans and vanilla extract. Mix thoroughly. Spoon into the bundt pan. Bake until a toothpick inserted into the center of the cake comes out clean, about 1 hour. To serve, sprinkle the warm cake with confectioners' sugar.

Makes 1 bundt cake

INDEX